AF606888

THE FURTHER READING LIBRARY

LOÏE FULLER
LECTURE ON RADIUM

THOMAS WILFRED
CLAVILUX AND LUMIA HOME MODELS

RICHARD SHARPE SHAVER
SOME STONES ARE ANCIENT BOOKS

RICHARD FOREMAN
NO TITLE

MARGARET WATTS HUGHES
SOUND MAY BE SEEN

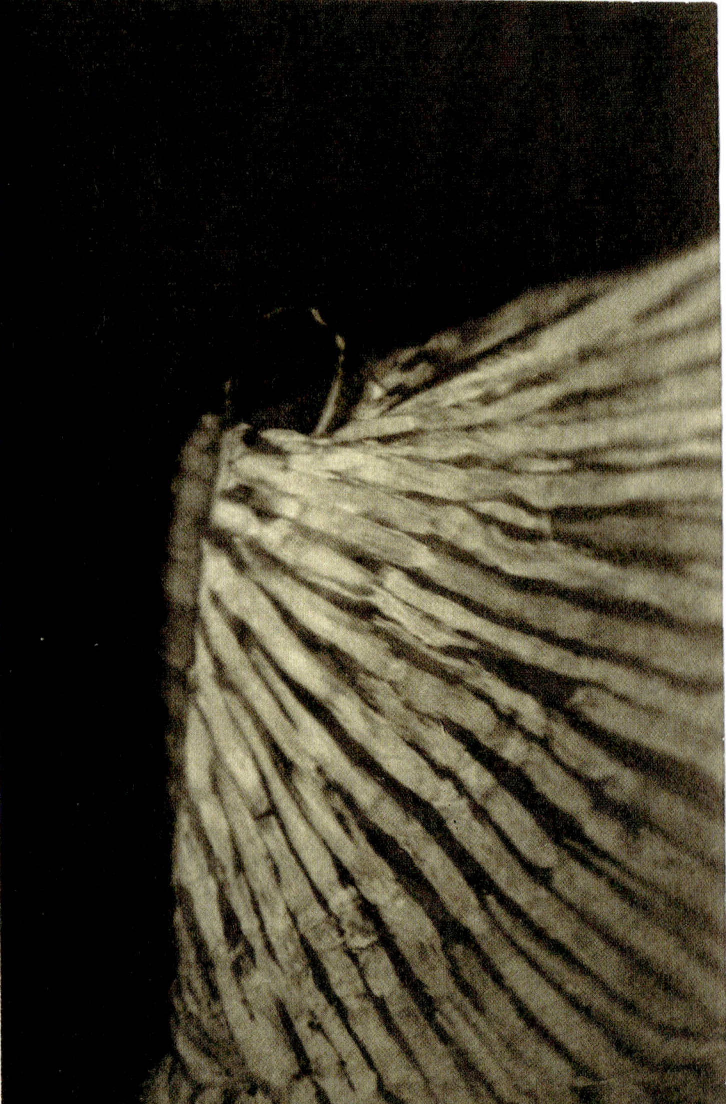

LOÏE FULLER

LECTURE ON RADIUM

THE FURTHER READING LIBRARY

CHRISTINE BURGIN BOOKS

SERIES EDITORS: CHRISTINE BURGIN AND ANDREW LAMPERT

Published by Christine Burgin Books
Ghent, New York

ISBN: 978-0-9976456-2-0

Published in 2025

www.furtherreadinglibrary.com

Designed and typeset in Caslon by Laura Lindgren
Image preparation by Jason Burch
Front cover lettering by Craig Pettman
Printed on 115 gsm Kasadaka White
Printed in China

FRONTISPIECE: *Loïe Fuller in her luminescent costume for* The Radium Dance, *ca. 1904*

CONTENTS

LOÏE FULLER AND THE GLOW OF DYING LIGHT 7
Tom Gunning

KAISER KRONE THEATRE PROGRAM 17
April 23–29, 1906. Fuller's handmade maquette for a weeklong program of performances including "her newest scientific creation The Radium Dance" at the Kaiser Krone Theatre, Kiel, Germany.

HOW I BECAME INTERESTED IN RADIO-ACTIVE MATTER 29
ca. 1911. Notes on meetings with Thomas Edison and Monsieur and Madame Curie; the history of her search for a luminous paint and the triumph and ultimate downfall of her Radium Dance.

LECTURE ON RADIUM 41
January 20, 1911. Draft of a lecture delivered at the Little Theatre in London, England.

LETTER TO CAPTAIN JOSEPH REINACH 59
November 1914. A proposal for using light to protect Paris from air attacks during World War I.

PRELUDE TO LIGHT 71
ca. 1911. An impressionistic meditation on light and color.

PATENTS BY LOÏE FULLER 77
1894–95

FURTHER READING 87

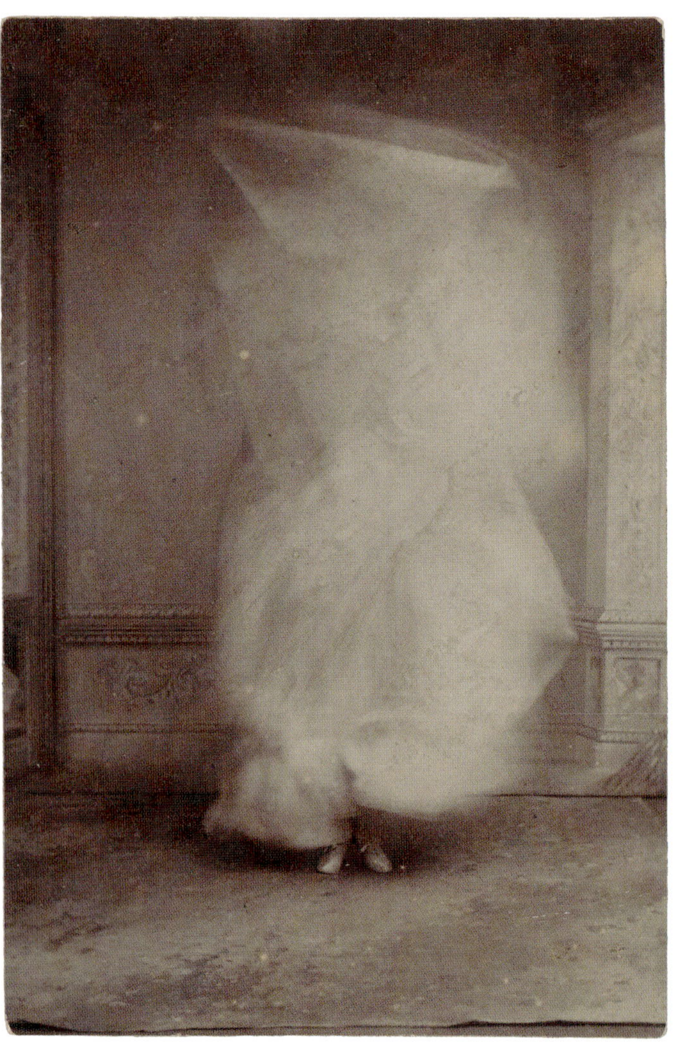

LOÏE FULLER AND THE GLOW OF DYING LIGHT

Tom Gunning

for Nell Andrew

Loïe Fuller danced with light as her primal partner. Her signature Serpentine dance was solitary, in the sense that she was the only dancer appearing on stage. But she was hardly isolated there; robes flowed about her as a form of kinetic sculpture, a virtual partner (in some stagings carefully placed mirrors multiplied her image, so she danced accompanied by her virtual selves). As she danced, she emanated her own environment, an ever changing envelope of shape and form. As Stéphane Mallarmé wrote of Fuller, "All emotion comes out of you and expands into a milieu."[1] This flow of shapes danced along with her; every movement traced an instant of creation, succeeded by another . . . and another—like the protean patterns of ocean waves, leaves rustling in the breeze, or the never-fixed flickering of a flame. These varied manifestations mesmerized Fuller's viewers. Her stage became a realm of contesting forces, a chaos that she evoked, then tamed, by the enchantment of her dance.

This flux of material, motivated and shaped by her moving body, ultimately became a succession of screens. As powerful

Loïe Fuller dancing, n.d.

as the forms that Fuller generated through manipulating her veils could be, light created another, even more ethereal, spectacle. A battery of magic lanterns supplied with carefully colored slides focused their beams on Fuller and dyed the figures she created with a variety of tints and sometimes even images: moons, stars, butterflies. Light supplied a new dimension to Fuller's constantly changing spectacle as it irradiated the dancer. No mere spotlight highlighting a star performer, this multihued projection generated a new visual delight. Fuller metamorphosed from a physically present performer into a virtual image—a dancing screen.

For Mallarmé, Fuller's dance achieved something he sought in his quest for a poetry that avoided direct reference: "To name an object is largely to destroy poetic enjoyment, which comes from gradual divination. The ideal is to suggest the object."[2] He wrote in a famous passage (anticipating Fuller): ". . . the dancer is not a woman dancing, for these juxtaposed reasons: that she is not a woman, but a metaphor summing up one of the elementary aspects of our form: knife, goblet, flower, etc., and that she is not dancing but suggesting, through the miracle of bends and leaps, a kind of corporal writing."[3] Mallarmé's Hegelian vocabulary celebrating the Idea tends to erase Fuller's physical power. If Fuller aspired to a novel concept of form, it should never be reduced simply to the immaterial; rather, her dance reflects the new understanding of matter in the modern era. Exploring the nature of matter and its appearance, Fuller's dance aspired not only to metaphysical demonstration, but to scientific experiment.

Fuller flourished in an era when the nature of reality was being radically reenvisioned. Fuller's parents were Spiritualists, and in her youth she attended a Spiritualist school, the Chicago

Progressive Lyceum. Nineteenth-century American Spiritualism attempted to found belief in immortality on empirical evidence and scientific speculation through apparent manifestations of the dead. Giovanni Lista believes Fuller's first performance, which suggested the Serpentine dance, may have been intended to portray a Spiritualist seance. He observed, "The dances of Loïe Fuller maintained a relation, simultaneously latent and manifest, with the imagery of Spiritualism and mediumistic seances."[4] Fuller pursued metaphysical investigations through her own medium. Her manipulations of light carried intimations about the protean nature of reality that led to her deep curiosity about, and experiments with, the new element radium.

Nineteenth-century scientific discoveries may seem to have widened the breach between a rational understanding of the world and traditional religious conceptions. But toward the end of the nineteenth century new discoveries seemed to offer a synthesis of the metaphysical and the scientific. Faraday's and Maxwell's demonstrations that magnetism and electricity were substantially identical; Roentgen's discovery of X-rays; and Hertz's radio waves—all revealed powerful invisible forces. This modern universe no longer resembled the Cartesian and Newtonian cosmos ruled by the physical collision of resilient bodies posited by classical mechanics. X-rays particularly exemplified the uncanny nature of these apparently immaterial forces: a ray of light that penetrated flesh to reveal bone. Not only did the X-ray offer a new form of vision; it raised questions about the nature and solidity of matter. Fuller called this new dynamic "Matterforce."

Speculation about this new model of reality suffused everyday life and inspired popular culture of the turn of the century.

The famous 1907 "locked door" puzzler *The Mystery of the Yellow Room* by Gaston Leroux (who wrote the even more famous thriller *The Phantom of the Opera*) initially explained a murderer escaping from a locked room by a scientific theory developed by Strangeson, the father of the novel's victim. Strangeson's "The Disassociation of Matter by Electric Action" claimed the existence of "fundamental properties of substance intermediary between ponderable matter and imponderable ether."[5] The theory, the novel claimed, had opened the way for the Curies' discovery of radium, and "was destined to overthrow from its base the whole of official science." Fuller subscribed to the belief that science had been fundamentally transformed by new discoveries and was gradually bringing to light things that had been deemed superstition. She wrote:

> Those things which are looked upon as unreal—supernatural—are one by one proving to be the results of natural causes, all part and parcel of this world and its products, the most advanced "unreal" fact being the telegraph without wires.[6]

Fuller's fascination with the newly discovered element of radium was triggered by the uncanny light it gave off when Thomas Edison demonstrated the X-ray to her. Edison showed Fuller the bones of her hand as revealed by X-rays, but her response was unique: "It was not the skeleton of my hand which interested me, it was the light and I wanted to know all about it."[7] Fuller immediately intuited that radium offered a different illumination from the electric light that she had projected onto her dancing figure. Electric light had itself been a novel technology when Fuller's performances began. Nell Andrew reminds us, "Fuller was developing her serpentine dance just as public electric power was spreading across Europe."[8] She called the new mysterious glow from radium "dead light" and "light's decomposition." She contrasted it with the electricity she had worked with.

> Radium is a soft light without rays
> Electricity is a hard light with rays
> Radium is a uniform steady light
> Electricity is a vibrating unsteady light
> Radium is a permeating light
> Electricity is a penetrating light.

She also claimed, "Radium is produced by unbuilding everything."[9]

Radium's unique light came from its decay. "Its light is a dead phosphorescence, half transparent, as a feeble light might look behind a ground glass. There are no visible rays."[10] This claim can seem contradictory, since the very name "radium," as Wikipedia states, derives from rays: "The naming of radium dates to about 1899, from the French word *radium*, formed in Modern Latin from *radius* (ray): this was in recognition of radium's power of emitting energy in the form of rays." But the "rayless" light Fuller described evokes the phenomenal difference between projected light and a glow emanating from material. As she saw it, this glow captured the death of light rather than its kindling.

For Fuller, the light emanating from radium dwelt within—and emerged from—darkness, rather than driving darkness away with focused rays. Soft, rather than hard; permeating, rather than penetrating (hence her attention in Edison's lab to the glowing light rather than the bones of the hand). Radium, Fuller claimed, released an energy that could upend all scientific principles, including gravity, and especially the previous building block of scientific conceptions of matter: the atom. Radioactivity derived from the breakdown of atoms emitting ionizing radiation to create phosphorescence, the new light, that so excited Fuller. She also underscored the element's radical nature through the gender of its discoverer: "the greatest of these discovered facts is due to a woman. Of course I refer to Madame Curie and Radio-Active Matter."[11] Radioactivity, as Fuller understood Curie's discovery, revealed that "atoms were always disintegrating" and in "perpetual motion" rather than supplying the ultimate elements of stability:

> atoms are in reality not atoms at all but intense, concentrated energy, material, shooting, darting, agitated, substance declaring itself as force or forces.[12]

Fuller intuitively sensed the energy that could be released from this unstable atom.

The actual use of radium for Fuller's costumes presented her with technical problems. The danger and expense of the element forbade its use, and Fuller turned to luminescent chemical salts for the glow she sought to reproduce. Yet her speculations on the significance of radium not only merged science and aesthetics but initiated an art suited to a new age of energy. Clara Nizard's pioneering dissertation terms Fuller's work with radium as "proto-nuclear aesthetics."[13] The "rayless" glow produced by radium had a very different relation to the light's source than her previous highly focused projected electric light. Nizard states this contrast with precision: "Here, all light emanates from the dancing fabric itself. Instead of replicating the flash of electricity, perception is organized by its attunement to darkness, a gradually strengthening glow."[14] Fuller's electric dances depended on light that her electrical technicians projected onto her: a mobile choreographic surface. But with the glowing costumes Fuller conceived for what she called her "radium dances," a soft glow would emanate from her figure and veils within a nurturing, surrounding darkness.

Nizard insightfully picks an emblem for this light from Fuller's repertory of images: "Radium took on distinct form: that of the moth."[15] The moth dwells in the night but is drawn to light, an attraction that often spells its death. We could christen this uncanny illumination with a term coined by another master of visual abstraction, the filmmaker Stan Brakhage, in the title of his most perfect film, *Mothlight*. Light as companion of darkness—rather than its conqueror—formed an aspect of the modernist fascination for "Artificial Darkness," which Noam Elcott traces in his magisterial study of the same name.[16]

As described in the *London Daily Telegraph*, early witnesses to first performances of her Radium Dances (achieved with phosphorescence rather than potentially fatal radium) recognized a kinship to darkness: "In complete darkness only the portions of material thus rendered luminous are visible. Hence the extraordinary ghostly effects."[17] As the *Telegraph* further reported, this performance included a "monster glowing moth, with shining antennae a foot long, eyes which are globes of light, and wings six feet high, glittering with luminous scrolls in all colours."[18] This spectral presence evokes Brakhage's description of his film: "What a moth might see from birth to death if black were white and white were black."[19]

Fuller had been warned by Mme. Curie of the dangers of this new element, whose radiation caused death and injury to scientists and workers in the early twentieth century unaware of its peril. This threat supports Nizard's intriguing claims that Fuller through her radium dances was a pioneer in "patterning the promise and terror of living with dangerous materials." Yet for Fuller it was the promise that emanated from this danger that inspired her work. Her explorations of the aesthetic potential of this new energy more than a century ago now haunts us with both that promise and that threat, its beauty and danger.

NOTES

1. Stéphane Mallarmé, "Another Study of Dance: The Fundamentals of Ballet," in *Divagations*, trans. Barbara Johnson (Harvard University Press, 2007), 137.
2. Stéphane Mallarmé, *Selected Prose Poems, Essays and Letters*, trans. Bradford Cook (Johns Hopkins Press, 1956), 21.
3. Stéphane Mallarmé, "Ballets," in *Divagations*, 130.
4. Giovanni Lista, *Loïe Fuller, Danseuse de la Belle Epoque* (Editions Stock, 1994), my translation, 76.
5. Gaston Leroux, *The Mystery of the Yellow Room* (Dover, 1977), 139.
6. Loïe Fuller, "Lecture on Radium," this volume, 43.
7. Fuller, "How I Became Interested in Radio-Active Matter," this volume, 29.
8. Nell Andrew, *Moving Modernism: The Urge to Abstraction in Painting, Dance, Cinema* (Oxford University Press, 2020), 12.
9. Fuller, "Lecture," 51.
10. Ibid., 47.
11. Fuller, "Lecture," 43.
12. Ibid., 52.
13. Clara Nizard, "Kinaesthetic Modernisms: Dance and Epistemology 1890–1960" (PhD diss., University of Chicago, forthcoming).

14. Ibid.
15. Ibid.
16. Noam Elcott, *Artificial Darkness: An Obscure History of Modern Art and Media* (University of Chicago Press, 2016).
17. "Radium Dances," this volume, 39.
18. Ibid.
19. The Film-Makers' Cooperative description of Stan Brakhage's film *Mothlight*, https://film-makerscoop.com/catalogue/stan-brakhage-mothlight.

Tom Gunning is Professor Emeritus in the Department of Cinema and Media Studies at the University of Chicago. He is the author of *D. W. Griffith and the Origins of American Narrative Film* (University of Illinois Press, 1986) and *The Films of Fritz Lang: Allegories of Vision and Modernity* (British Film Institute, 2000), and over a hundred and fifty essays on cinema and related arts, including two on Loïe Fuller and cinema.

KAISER KRONE THEATRE PROGRAM

Fuller's handmade maquette

April 23–29, 1906

The radium is a soft strange light
It is like the moon
It throws no rays
It is not brilliant
And it must be seen in absolute darkness
It is a new unknown light

LA

Loïe Fuller

FOR

THE FIRST TIME IN KIEL

One week

ONLY

APRIL 23.

KAISER KRONE

PROGRAMME
NO I

MONDAY. TUESDAY. WEDNESDAY. THURS
APRIL 23. 24. 25. 26

LA
LOÏE FULLER
WILL APPEAR
FOR THE FIRST TIME IN KIEL
IN HER
LATEST CREATION
OF
LIGHT COLOUR
AND
MOVEMENT
THE DANCE
OF
1000 VEILS
WITH HER TROUPE OF LADIES
AND
CORPS OF ELECTRICAL
ENGINEERS
AND
SPECIAL STAGE DECORATIONS
OF
WHITE GAUZES

TABLEAUX.
I

THE SNOW STORM
THE SEA
THE SHIP
THE TYPHOON
THE HEAVENS
THE MOON
THE ARCTIC REGIONS
THE NORTH POLE
THE PASSING OF THE SOULS
AND
DANCE NO 1.

TABLEAUX
2

THE ROLLING CLOUDS
THE LIGHTNING
THE FLASHING LIGHTS
THE VOLCANOES
THE SWEEPING FIRES
GLIMPSES OF HADES
AND
DANCE NO 2

TABLEAUX
NO 3.

SPACE
FALLING STARS
THE MOON
THE BLUE OF ETHER
THE HOVERING ANGELS
THE MORNING RAINBOW
THE
FLYING = LIVING
BUTTERFLIES
THE FORESTS
THE STAR OF NIGHT
THE SETTING SUN
AND
DANCE NO 3.

PROGRAMME
NO 2
FRIDAY SATURDAY SUNDAY
APRIL 27. 28. 29.
LA
LOÏE FULLER.
WILL APPEAR
FOR THE FIRST TIME IN KIEL
IN HER
ORIGINAL SUCCESSES
THE FIRMAMENT
THE FIRE
THE GREAT WHITE LILY
MANIPULATING A DRESS OF 500 YARDS.
AND
THE DANCE WHICH MADE HER
FAMOUS
THE SERPENTINE
AND
HER NEWEST SCIENTIFIC ~~DANCE~~ CREATION
THE
RADIUM DANC

PROGRAMME

FRIDAY SATURDAY

AND

SUNDAY

ABSOLUTELY

LAST DAYS

OF

LA LOIE FULLER.

1. THE RADIUM

2 THE SERPENTINE

3 THE FIRMANENT

4 THE FIRE

5 THE LILY

IN

SPECIAL BLACK

DECORATIONS.

Note. The RADIUM is A SOFT STRANGE LIGHT

It is LIKE THE MOON.

It THROWS NO RAYS

It is NOT BRILLIANT

AND it must Be seen in ABSOLUTE DARKNESS

It is A NEW UNKNOWN LIGHT

AND SHOULD NOT BE LOOKED UPON

OR JUDGED BY A BRIGHT LIGHT.

AND It SHOULD Be seen very NEAR.

RADIUM

There is no large quantity known
to exist in the world.
But in the reduced or
concentrated residues of Pitch blend
And the Strontium group of
phosphorescent substances
known as Barium, calcium
Etc etc contain minute
quantities of Radium.
It is these substances
which Miss Loie Fuller
employes to permeate a
robe so that it may in
the dark appear in its own
light. The above named
substances are in their
condensed state rare
and costly. Only about
10 or 12 grams being
made in 24 hours in
the greatest chemical
laboratories. Therefore to

enable her to produce a
sufficient quantity wit which
to permeate a robe much
time & money were expended
by Miss Fuller who when
she began ten years
ago to study these materials
which give forth light
she did not know whether
she ~~[illegible]~~ would ever
succeed in accomplishing
her desires. = The attaching
of the tiny crystal salts to the
material requiring an
adhesive substance which
melted & dispersed the
powder robbing it of half its
qualities of light. Thus
adulterating instead
of leaving it concentrated!
The cost of experiments —
amounted
to from 2 to 3000 marks

for each
Kilo obtained, the sulphur of Zinc being double that amount ~~& impossible to produce in large enough quantities for stage purposes~~, the Radium light dance of tonight is from the Strontium group of salts in which minute quantities of Radium exist. Hence the reason of the ~~title~~ Radium dance. The Radium itself looks in the dark like a piece of decayed fish & not at all remarkable, but it has the faculty of rendering other things extremely luminous

And of effecting materials diseases & assuming electric currents.

For instance

In the dark a morsel of Radium may be concealed and hidden, but instantly a diamond any where in the vicinity of the tube which contains the Radium becomes luminous to its highest degree.

and after 20 or 30 minutes a <u>person</u>, clothing will become phosphoro

& yet the Radium may not be visible.

also

touch for a few moments the bare flesh with the tube containing the Radium

And although no
sensation whatever is felt
a sore will break out 3 or
four months later which
is almost incurable!
and
Two needles held up
horizontally in the air
by electricity will drop on
the approach of the
Radium tube, thus
annulling the power of
electricity. No one can
explain these things.
Few facts are known about
the Radium, but even the
materials in which it is
found are interesting &
these materials are the
light Miss Loie Fuller
will show in ~~[illegible]~~ in
her so called
Radium Dance.

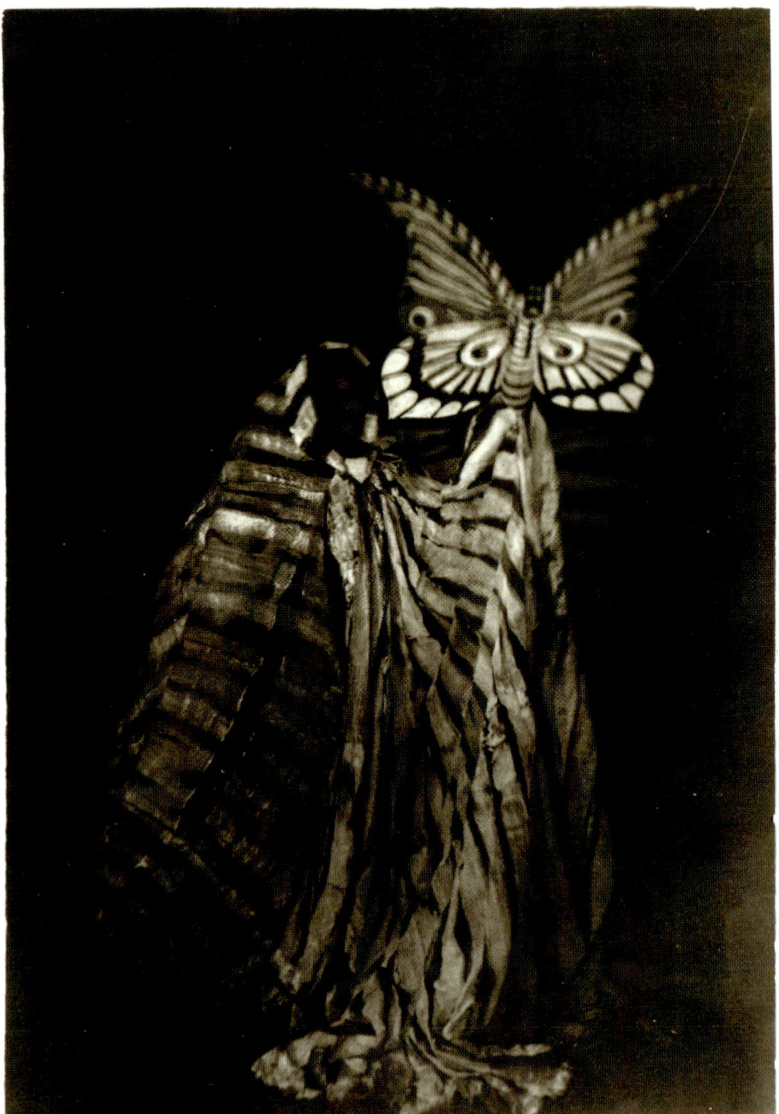

HOW I BECAME INTERESTED IN RADIO-ACTIVE MATTER

ca. 1911

I thought what a surprise that would be, I who am deluged in light, dancing without any light at all. That it was science never occurred to me, for I had a workshop of my own which was in part what I now know to be a chemical laboratory . . .

This text appears in a notebook immediately following a handwritten draft of Fuller's "Lecture on Radium." The location and opening line suggest that this reminiscence might have been part of her lecture. The autobiographical material it contains—Fuller's descriptions of her relationship with the scientists of her day and of her own experiments in search of luminescence—is not included in her 1913 memoir Fifteen Years of a Dancer's Life: With Some Account of Her Distinguished Friends.

PAGE 28: *Loïe Fuller in her luminescent costume for* The Butterfly Dance, *ca. 1904*

And now I will explain how I became interested in Radio-active matter.

In one of my visits to N.Y., I was taken to see Mr Edison and there for the first time I saw a chemical laboratory! It was exclusively used for those researches which dealt with Radio-active matter. I went there interested in electricity and came away filled with the desire to know more about that wonderful light produced by phosphorescent salts.

In Mr Edison's laboratory there were several chemists preparing these salts out of acids. Little glass bowls were arranged on long tables and shelves and the men were dipping out the crystals as they formed in the bottom of these little vessels. The process of this drying up of liquids was a long and tedious one. Mr Edison explained to me that many of these crystals produced colour in their light and his researches were being made to find a combination of acids that would produce colourless crystals, because colour was a great element against transparency and transparency was the chief aim in the use of the salts for the X Ray. I may note here that Mr Edison afterwards gave up this part of his work, the acids and X Ray causing such sad results to the men in the execution of their work. Of course this is now well known.

Mr Edison took me to the dark room to see the X Ray. For the benefit of those who have not seen it I must explain what the thing that I saw looked like. It was a little box just like those in which we see stereoscopic views only there was a top over it, so it was all closed in. You looked in it just as you would if you were going to see the pictures. Only inside of it, instead of a picture you saw a little wall of light. Outside of this box you held your hand at the back against the wall inside of which the pictures would have been put. Then back of your hand was held a tiny electric bulb with a small greenish light in it. Until this light was there the inside of the box was darkness! As soon as this light was turned on, I saw all the bones of my hand, and the flesh about them looked like a veil. Then Mr Edison explained to me that the wall in the box was covered with phosphorescent salts, and these salts consumed the light as sand did water, and that was why one could see the bones; they were outlined against the light because they were thick and solid, and the flesh around the bones resembled matter like a veil because the light filtered through the fingers as water would if you pressed your hand down on wet sand.

This curious stuff all aglow with a light I had never seen before held me spell bound. It was not the skeleton of my hand which interested me, it was the light and I wanted to know all about it. Then Mr Edison explained it technically and scientifically to me. I need not go into this part, it is so well known and to those who do not know it technical terms mean nothing. I should have to explain to you about chloride of Barium—the power of the Ray X, etc. etc.

While we were examining this little X Ray box it suddenly occurred to me that if I could have a dress permeated with that substance it would be wonderful but how could I dance with an X ray lamp behind me to set the dress permeated with crystal alight and I asked Mr Edison if the salts would retain their luminousness when the lamp was gone.

He hadn't thought of that—so we tried it, and lo and behold the light remained one or two minutes before it began to fade away but the light never quite went out; something kept them feebly illuminated all the time only one could not observe it until they had held their eyes to the chamber for a long time. Once the little daylight which filtered through into the dark room had entered the sight one could only see darkness in the little box, till the X Ray was turned on again. I discovered this little constant illumination through watching the thing fade away—after half an hour it was still there. I called Mr Edison's attention to it, but there was no explanation to be found—and we thought no more of the matter. Since then Radium and other active matter having been found in traces where these phosphorescent substances abound, their continual light is evidently due to the effect of these traces, which possess the faculty of illuminating everything in their presence.

But to return to Mr Edison. I explained to him what I wanted to do, and he consented to see what result he could obtain, so I left a large gauze scarf with him to experiment upon. I never knew these results, because I was away from America for a number of years and when I had an opportunity to again see Mr Edison he had as I say given up his chemical experiments.

I was still interested—in fact more than ever, to have Radio active dresses to dance in. I thought what a surprise that would be, I who am deluged in light, dancing without any light at all. That it was science never occurred to me, for I had a workshop of my own which was in part what I now know to be a chemical laboratory for there I and my assistants not only dealt with and used numberless scientific electrical machines of enormous power for producing great rays of light, but I was occupied with studying through the microscope infinite colours, and experimenting with chemicals to reproduce them, for my coloured and colour effects lights.—I never knew that was science either, I just thought it was work to bring about the result I wanted! So the making of phosphorescent salts seemed only different, that was all.

Well, at the house of Camille Flammarion I met Mon D. L'Oncle the French Deputy and learned that he was a great optician and chemist. Delighted at this, I told him about my experience with Mr Edison and he proposed to help me. Soon after that we went to work with a will till one day we had an explosion—and that stopped operations for a time. We had prepared a huge veil and dress with a coat of sulphur of calcium. We burned a calcium ribbon with which to illuminate it when suddenly there was an explosion and we knew nothing any of us for some time. The place caught fire and we nearly all burned up. My eye brows and eye lashes and all the front part of my hair disappeared too. So my enthusiasm had a set back for a time. Later on it all returned however.

A French sculptor friend called to see me one day with an old gentleman who turned out to be a very great professor of chemistry and that brought the subject of my phosphorescence up again. This distinguished old gentleman took a deep interest in it, and we went to work this time in his laboratory. Before long we had a few little grains of sulphur of calcium, sulphur of zinc, etc producing violet, green, yellow and red—sulphur of calcium producing violet, sulphur of zinc green etc etc. I was in ecstasy. The next thing was to find the way to apply it to silk, and here we found our great stumbling block. The liquid gum melted the stuff before we could get it stuck onto the silk and when this was accomplished half the powder was lost. This was rather serious in the sulphur of zinc for it cost over 100 Francs for half a pound. The others were less expensive. And as my dresses were voluminous we required a great deal more than half a pound. Also, when we had accomplished the application of the salts to the dress, it was like a board and nothing could be done with it. I couldn't even get it on. It was very discouraging to say the least, not to mention all the money spent and we had been at it then over two years. At last ingenuity filled the place that science couldn't find.

I observed that cloth with only patches on it of salts became all illuminated in the dark when it was in motion. So it occurred to me to put the stuff on in stripes thus leaving the full skirt to hang in pleats and folds. We had found the suppleness we had worked so long for in the wrong lines, the lines of thinning the gum and substance, till neither were any good.

Then we took a great dress which contained 100 yards of silk and went to work all over again. It was the biggest dress I had, we used the least expensive substance (sulphur of calcium) on it and this was expensive enough, for we used over six pounds in weight in the operation.

Then we made a dress and a veil out of black gauze spotted with violet. This was exquisite. Thrown up into the air, the black veil disappeared in the darkness and only the falling luminous drops were seen, which became elongated in the descent, taking on the form of luminous tears; the effect made me seem veiled in tears, the black gauze of the dress being invisible, only the large luminious drops falling around me. Then we made great butterflies of red, green and yellow. They were marvelous. Like huge spirit things, so luminous and ethereal they were, one felt in touch with the supernatural, now explained.* The impression given was that I seemed so far away from the dress and I could be seen from head to foot but as if I were a black statue.

It was at this time that I met Madame Curie, and nothing could be more curious or interesting but that does not enter this lecture. It led up, however, to a visit from Mr and Mrs Curie and all the professors of the Sorbonne to see the results of what I had done with Prof Janin. By this time I had a chemical laboratory again of my own and it was here they all came to see me. We had a sceance one evening that lasted far into the night. Some of my friends who joined the party thinking to render me justice reported it all in the *London Daily Telegraph* and that paper came out with an article a column long headed in big lines Loïe Fuller Radium Dancer. It was just then that I had to go to So. America to appear in those dances with which I have been associated and I left my luminous dresses with my dear old Prof. to await my return.

When I got back 6 months later I found that that article in the *Telegraph* had given birth to so called Radium dances all

*In a second draft of this text, Fuller adds: "We took the big dress to Sarah Bernhardt's enormous theatre and when it was all in motion it lighted the Auditorium up so that you could read in the farthest corner."

over the northern continent. They had begun about 6 weeks after that article had appeared, so, when I offered my invention I was told that Radium dances were played out and nobody wanted them. The name had been used up on something that was as far from my invention as a candle is to an arc light! It was luminous paint!* And so my dance of tears, of spirit butterflies, of green crawling leaping ghosts of serpents and the great silver moon have never been seen. Instead of that I am here telling you about them and how they led me into the land of Radium, because it was at this time I came into contact with it through Madame Curie, and as I have said nothing could be more interesting, but I did not study with her, notwithstanding a statement which has been made to that effect. Indeed I would be proud to have had that great honour but the statement is not true. Although what little I know about Radium I learned through that marvelously gifted woman and about whom I may perhaps say this much. No woman has ever lived whose unpretentiousness and simplicity are so remarkable that no success no adulation no fortune can be great enough to tempt her from her simple faith or spoil her. The very details of her life and work would give you wondrous and overwhelming evidence of this, if I were at liberty to describe them to you._

*In a second draft of this text, Fuller noted that the luminous paint used by her imitators was: ". . . the crudest and very lowest form of phosphorescent paint! I had used the phosphorescence with a reason, to express something."

PARIS DAY BY DAY.

RADIUM DANCES.

BY SPECIAL WIRE.

From Our Own Correspondent.

PARIS, Tuesday Night.

Court Circular.

BUCKINGHAM PALACE, March 1.

His Majesty the King, attended by Lord Lawrence, Major-General Sir Stanley Clarke, and Colonel A. Davidson (Lord and Equerries in Waiting), left the Palace this morning for Cambridge, to inaugurate the new buildings of the University.

HIS MAJESTY'S THEATRE.

THE WEATHER IN FRANCE.

From Our Own Correspondent.

FASHIONABLE MARRIAGE.

LAST OF THE LYCEUM.

SALE OF RELICS.

AN UNFORTUNATE AFFAIR.

RICHTER CONCERTS.

CLERICAL ATTIRE.

LORD ROSEBERY ON WALES

ST. DAVID'S DAY DINNER.

THE DAILY TELEGRAPH, WEDNESDAY, MARCH 2, 1904

Paris Day by Day. Radium Dances.
By special wire. From Our Own Correspondent.
Paris, Tuesday Night.

Loïe Fuller, who is nothing if not modern, has just invented "radium dances," which she has shown to M. and Madame Curie. The famous physicists said that the American serpentine lady had made them look upon their own discovery in a new light. The name "radium dances" is not, as might be imagined, a fanciful one. No actual salt of radium is manipulated by the lady in the production of her new effects. But she uses substances very nearly akin to radium. These are certain fluorescent salts, extracted from the residue of pitchblende, whence the Curies obtained the mysterious matter discovered by them. It has, in fact, been supposed, as will be remembered, that such salts derive their fluorescence from the influence of radium. So much for the justification of the name given to the new dances. The spectacle itself is weird and fantastic in the extreme. The few guests asked by Loïe Fuller to witness in her studio the performance, which has not yet been given in public, are marshalled at one end of the gallery, with all lights put out. Through a slit in the curtains opposite a green glow is seen. Suddenly an apparition comes into view. It is a vague form, only distinguished by the hundreds of tiny glow-worms which it seems to carry on its flowing raiment. The tissue of twinkling stars floats about, circles, sweeps along the floor, or is wafted up until it is shaped into a sort of great luminous vase. The dancer's face is never seen, her form is vaguely divined when outlined by the glowing lights. The apparition vanishes, to be followed by another more weird still.

Above a perfectly invisible head, which you only suppose to be there, shines a bluish halo. Below, clothing an unseen figure, is a long robe, which is merely one great patch of the same ghostly light. The apparition slowly moves to a solemn rhythm, seems to invoke heaven, the halo being thrown backwards when the head is, as you conclude, uplifted, and, finally, the robe of light sinks on to the floor, when you infer that the figure kneels. The second ghost vanishing, a third appears, a monster glowing moth, with shining antennae a foot long, eyes which are globes of light, and wings six feet high, glittering with luminous scrolls in all colours. The moth flutters round and round the studio, then out of our sight, but reappears instantly, accompanied by a smaller, glowing white butterfly, which beats its wings over the monster luminous insect's head. Lamps being relit, we are brought back to reality out of ghostland, and, having an opportunity of examining the dancer's dresses, find that they are made of a peculiar kind of silk completely impregnated with certain fluorescent salts. In complete darkness ony the portions of material thus rendered luminous are visible. Hence the extraordinary ghostly effects. In the treatment of stuffs by fluorescent salts lies the originality of the invention. M. Curie, keenly interested in the idea, said that it was an application of the properties of radiant matter of which he had not thought.

Lecture on Radium

by

Loïe Fuller

Jan. 20 – 1911. – London England.

Science is, as we all know, the uncovering of facts in nature, and as nature discloses her truths, to us she opens the doors of understanding & reason. — Those things, which were looked upon as unreal & supernatural — are one by one, proving to be the results of natural causes, all part & parcel of this world & its products. The most advanced "unreal" fact being the telegraph without wires. = = On reflection these things bring us face to face with the greatest result of all progress the end of superstition, —

A great theologian once said to a great agnostic, "If you do not believe in a hereafter, you would not as of our

LECTURE ON RADIUM BY LOÏE FULLER
January 20, 1911, London, England

Alas, how often we have to unlearn those things we have learned and science is no exception to the rule.

Loïe Fuller delivered her "Lecture on Radium" at the newly opened Little Theatre in London on February 2, 1911. Two handwritten drafts of the lecture exist. The reproduction on page 40 comes from a draft dated January 20, 1911, which includes a lengthy digression on the history of science, focusing on the initial rejection of the ideas that later prove central to understanding the universe. The second, undated draft, which is transcribed on pages 43–55, is a substantial revision that reduces historical references and culminates with Professor Crookes' Spinthariscope, a device that makes the existence of atoms visible to the naked eye.

Science is—as we all know—the uncovering of facts in nature. As nature discloses her truths to us, she opens the doors of understanding and reason!

Those things which are looked upon as unreal—supernatural—are one by one proving to be the results of natural causes, all part and parcel of this world and its products, the most advanced "unreal" fact being the telegraph without wires.

On reflection these things bring us face to face with the greatest result of all progress, the end of superstition.

A theologian once said to an agnostic, "If you do not believe in a hereafter you would rob us of our immortality." "If it's a fact, I couldn't," he replied.

And so, if unseen facts in nature shall become real, there is no longer reason for superstition to exist. Its death will be as natural as its life has been. For which fact we are indebted to science, to those men and women who devote their lives, for the benefit of mankind, without any material gain to themselves to speak of other than the results they obtain! May I be excused for taking pride in calling to mind that the greatest of these discovered facts is due to a woman. Of course I refer to Madame Curie and Radio-Active Matter.

What little is known about these substances is generally pretty well known, so I cannot hope to electrify you with the announcement of new discoveries about it. But finding the subject of Radium such a powerful and interesting one, it occurred to me that there might be other people less fortunate than myself, others who lack the same opportunities for coming in contact with the subject, so this short and simple explanation is given with only the idea of interesting them and not instructing, technically or rhetorically, anybody.

I have always myself resented having learned expressions thrown at me, for it seems so much easier to call things by their simplest names. Who cares whether Radium is called Radio-Active Matter, an earth salt, an illuminated crystal or reduced Pitchblend. Naturally, by the same reason that we want to know how trees are made into furniture, there are some people who are not scientists who would like to know how Pitchblend is turned into Radium. For their benefit solely I will begin by quoting some of the things that have been recorded and which all students of this branch of science know.

The discovery of Radium was not—as discoveries usually are—the result of an accident, but it was an accident, insomuch as it was an unexpected result of a long and tedious chemical research on the part of Mr and Mrs Curie, the final result having been found by Madame Curie herself.

It was the result of separating and reducing matter to the smallest possible quantity.

It is well known that certain materials exist, which, when exposed to a strong light (artificial or daylight) become luminous, and Madame Curie has found where this luminous or phosphorescent earth salt abounds, that there existed sister elements in minute quantities which are always illuminated without ever coming into contact with any light. And more, they have the quality of making everything in their presence more luminous by far than they are themselves, although in volume they are only microscopical.

These phosphorescent materials or salts are found in Pitchblend.

Pitchblend is earth which contains Uranium (I may say here that Uranium is that substance inside the electrical bulbs which are used everywhere).

As I say Pitchblend consists of Uranium, Lead, Iron, Barium, Calcium, etc. and there are in this Pitchblend traces so very minute they are called elements and Radium is about one 16th hundredth part of these traces. To obtain this minute part is to first extract the Uranium, which forms over 80 percent of the 100th part, and the remaining 20 percent of Lead, Iron, Barium, Calcium, and traces that fuse with carbonate of soda, dissolve in Hydrochloric acid. Precipitate the lead and other metals by the aid of Sulphurated Hydrogen. This treatment removes the metals and lead. What remains are the chlorides of Radio-Active Matter. The earliest crystals recovered are Radium. The others, Polonium, Actinium, etc. are infinitely more rare. Pitchblend, after having been treated for the extraction of Uranium, is called refuse or residue. It looks like a reddish and lumpy powder.

In 8 tons of this residue was found one gram of Radium, not wholly pure at that and the rarer substances in such infinitesimal quantities that man cannot conceive them. For instance, one 15 hundredth part of a grain of Polonium was found in 2 tons of residue. Remember this refuse had already had extracted from it over 80 percent, so, add to these 2 tons of residue the 80th part, one can soon calculate that these two tons represented but 20 percent of the material in its original state. So one can see that tons and tons upon tons are required to be worked upon and reduced to nothing in order to be able to find these wonderful, little microscopical grains of sand as it were. All this indicates to us very plainly the vastness of the time and the amount of patience required in the work.

The first quality or extraordinary phenomena observed was: the first Radium procured by Madame Curie, she put in a glass tube then left it standing in a small measuring glass overnight. The next morning the glass had changed in colour to a deep violet. The tube containing the Radium was not affected!

Radium is kept in a tube of glass, one of quartz, one of metal, one of rubber and one of Gutta-percha, all in order to confine its powers, and to preserve its quantity, but science reckoned wrong; nothing <u>could</u> confine its powers, nothing <u>could</u> diminish its quantity.

In order to see Radium the outer tubes are taken off and one looks at it through a little square of transparent quartz, which is inserted in the tube just over the spot where the Radium is confined. Both ends of the tube are protected with rubber, in order to minimize as much as possible the danger of handling it.

It cannot be seen at all, until you have been in a very dark place for some 10 minutes—and then—even before it is uncovered, if there are diamonds in the room they will become brilliantly illuminated but imitation stones will not be affected at all.

The Curies soon found they were in possession of something unseen unseeable. Something which had to do with those forces hitherto unknown as material. Here was something that began to give proof of things.

Radium began to open up the laws of the Universe and show things that science had never been able to explain. It showed that science had based its calculations on some theories at least altogether wrong. It has shown us the disintegration of the atom, a fact which changes man's entire conception of matter and force and perhaps gravitation. Its existence, its manner of force, is in direct opposition to the law of cause and effect. Its power of light, lighting up things, its power of heating, of burning, all, all are contrary to nature's laws as we know them.

In the light, Radium resembles a tiny bit of table salt. In the dark it looks like a bit of old decaying fish. Its light is a dead phosphorescence, half transparent, as a feeble light might look behind a ground glass. There are no visible rays.

No one on seeing it would dream for a moment of the powers of this insignificant 10 or 12 inch tube with its little phosphorescent light.

But it is always, always illuminated, that is one of its marvels! From where does it get its light and from what? Nobody knows. It produces heat. From where? From what? Nobody knows. These evidences are contrary to the law of cause and effect. Science is baffled.

Ah, facts frequently want explanation to be facts. Indeed how seldom facts are facts without explanation.

On reflection this reminds us of how, in the ordinary things of life, we often explain our attitude by saying "I didn't know" and how very often we change our opinion of a case when it is explained to us! Although the fact itself may have had nothing to do with it.

Alas, how often we have to unlearn those things we have learned and science is no exception to this rule.

Radium, that magic word Radium. But what magic? Let us first understand in what way we mean to use that word. Magic, magic is a mystery and we call a thing a mystery because we do not understand it. There are two magics and many mysteries which are not magical, but I wish to refer now to those which come under the head of magic.

There are two kinds of magics, I say magics in order to simplify what I mean. One kind is created by man, things which are magical or mysterious to everybody but himself, because to him they are simple results due to natural causes manipulated by him.

Then there are nature's magics, called magics because no man understands them. As soon as their causes are known they are no longer mysteries, no longer magics. They are natural results of natural causes.

Radium is one of these natural magics because as yet no one understands, no one can explain its being, its cause of being nor its results of being. But the proof remains that it has revealed facts in nature hitherto unknown and undreamed of. No explanation can undo this.

What is Radium. A question very easy to ask, and I may say is just as easy to answer, I do not know. When it comes to be better known, there will be less reason to talk about it.

It is natural for us to discuss most those things we know the least about. Because each one by making known his views his opinions his little discoveries and observations, starts and pushes along a snowball in thought, which gathers force and material as it goes, till research and comprehension are apparently rewarded. The results are placed on record and man takes up something else upon which to concentrate his forces, his intelligence. Knowing that the structure he has built up today nature will undo tomorrow, but this is nature's very means of education in the developing of man's reason. It is in this spirit I am talking to you of the very few things I have heard, seen, observed and think about Radium.

If all the analytical chemists who have treated Pitchblend for Uranium and thrown the residue away could get it back again, we might soon have enough Radium to experiment with and learn more about it. But who knows, perhaps nature is wise in revealing it to us slowly. Perhaps it is well that Pitchblend does not abound anywhere.

One day a scientist in Paris had occasion to experiment with an electric machine which lifted into the air and held up horizontally thus two large brass needles or arms. As the professor approached the machine the needles fell powerless. When he receded they flew up again.

It was some time before he realized it was the Radium which he held in his hand. It had traversed its many tubes and annihilated electricity unseen through space, and it wasn't as big as the head of a pin. Nothing could confine the Radium. The next thing was for science to obtain more of it. What existed would soon throw itself or its force all away. But month after month passed, then year after year, and this tiny bit of force remained the same, the same tests were followed by the same results: no waste. Another opposition to our known laws.

This started science to calculating and figuring on its actual minute loss or use of force, and I think I remember rightly when I say that Mrs Curie is said to have calculated mathematically that Radium would lose one half of itself—half of its force, in 1700 years. Of course larger quantities might give a different proposition to science. Polonium, that sister element far more rare, on the contrary is said to disappear I think completely in 40 days. I have been told this, but I cannot recall whether it was by Madame Curie herself or not.

Let us dwell on the heat in Radium for a moment. A French scientist held the Radium tube to his wrist for an instant. He sensed, he felt nothing. A long time afterwards a red spot appeared and it got worse. The doctors could neither diagnose or cure it. The Radium had burned without burning. Another law defied!

Without the removal of its coverings or tubes Radium in the dark throws itself or its powers out through space and makes everything luminous around it.

If you approach the eye with the Radium (still in all its tubes) for a part of an instant only, the eye will be filled with a light so powerful, it is as if you had been looking at the sun, only it remains much longer. Yet the eye would have seen no light, the Radium not having been uncovered.

This is opposed to the law that the eye, like the photograph plate, must be exposed in order to receive an impression. Light might pass through the eyelids, but in the case of the Radium, all was absolute darkness. An experiment not safe to try often. This evidence is against the law of cause and effect as taught by science.

I ask myself, How is it that the Radium did not destroy, burn or illuminate its own tube coverings! I will ask this when I get an opportunity.

Radium will pass through the finest glass blown and render luminous phosphorescent crystals confined in a vacuum.

If Radium and its sister elements can bring to our own vision those things which we cannot see, as it does the atoms, its influence and importance in the world of progress cannot be measured. Indeed its wonders are already beyond our calculations or our understanding and the words "I'll believe when I see" are of no longer any value. Their argument is past, for now we know that the naked eye can see almost nothing at all. Materialists and science will have to find another phrase, and faith may become a fact.

Radium Versus Electricity
Radium is a soft light without rays
Electricity is a hard light with rays
Radium is a uniform steady light
Electricity is a vibrating unsteady light
Radium is a permeating light
Electricity is a penetrating light
Radium is a light that never goes out
Electricity is a light that always goes out
Radium is a light that depends on nobody knows what
Electricity is a light that depends on everybody knows what
Electricity is a great power
Radium is a greater power
Electricity is produced by building up something
Radium is produced by unbuilding everything
Electricity must be applied for effect
Radium must not be applied at all
Electricity is ruinous to the eyes as a light
Radium as a light is the reverse.

If sufficient quantities of Radium could be found and used for lighting purposes it might revolutionize our ideas and manner of using light and our very manner of living might change.

Phosphorescent light is so marvelous.

Through the medium of Radio-Active Matter we are learning—I repeat—that those laws based on theory and past knowledge which could never be explained are most of them wrong!

The theory (only theory) but called law of the atom is laid dead.
The theory (only theory) but called law of gravitation is a question.
The theory (only theory) but called law of cause and effect is defied.
And the theory—only theory—but called the law of the invisible or unseeable things is questioned, for everything may be visible.
Radium, the smallest thing in the world, has done all this.

And now, Professor Crookes, the eminent English scientist, has discovered something through the use of Radium of very great interest and of still greater importance. He has found that Radium can so illuminate certain substances that with the aid of a magnifying glass it can be seen that atoms are in reality not atoms at all but intense, concentrated energy, material, shooting, darting, agitated, substance declaring itself as force or forces.

Madame Curie has made the count of this energy or force and her calculation is said to be 2,000 kilometers per second (and this world whirls through space but 1,000 miles an hour). It is strange how man can make mathematical calculations which are far, far beyond his comprehension, a fact which might suggest that comprehension and brains are two distinct things.

But was it not marvelous for science to discover the atom? With 3,000,000 together the eye could not see them.

This disintegration of the atom can be better understood perhaps, when we reflect that it is said 3,000,000 atoms can lay side by side in one molecule of the air, and in turn each atom is itself bursting into millions of parts. These parts are called Electrons and are no longer material as science knows it.

Without Radium, science of today would know no more than science of the past that the atom was not the smallest thing in the world. Until now there was no way on earth to know it.

Prof Crookes invented a little instrument called the Spinthariscope in which can be seen a phosphorescent substance, Barium I think, spread over a small surface called a screen at the end of a tube and illuminated by an invisible bit of Radium held in front of the screen by a hand like the hand of a watch.

When we reflect on the smallness of the atom it might occur to us that we had to learn to conceive before such a daring thing could be shown or revealed to us as an atom bursting into 3,000,000 parts and this is what we see in Prof Crookes' Spinthariscope.

In this little instrument can be seen by the naked eye absolute proof of the falsity of the theory of the atom. Atoms are always disintegrating. The Radium only lights them up so the eye of man can see them.

This then means that everything we called an atom is many atoms in perpetual motion. The word atom only marks a condition of matter or, rather, force which is called matter!

It explains perpetual motion of everything, from the planet to the atom, from the atom to the electrons (or our last knowledge through Radium of matter!) and suggests that the law of gravitation is less apt to be holding us on the earth than the simple movement of whirling through space. Just as water is held in a bowl if the bowl is whirled around fast enough, a falling apple by the same movement might be drawn to the earth, who knows. As the Spinthariscope of Prof. Crookes gives evidence that matter is only force! There are not two things in matter there is only one: Force. To be understood—one word: Matterforce.

I would like to say that I am well aware no scientist will make any assertion with regard to a new discovery (no matter what he may believe) until he can produce proofs, for if he should be found in an error he would cease to be an authority on so and so. But the fact that he follows given lines shows that he is thinking about something he does not yet know and of which he will not speak. I often feel it would be worth much to know the reasoning, the impressions, the hopes and beliefs of a scientist. Not being a scientific authority myself, I feel at liberty to say what I think and what I may feel about a thing if my reason suggests it. But Radium and the Spinthariscope speak plainer than any theories have ever done. They declare facts!

We women should be proud that it was through one of our own sex that these wonderful findings of Prof Crookes were possible.

The law of miracles is opening up to us in the invisible, as it has for years been doing in the visible world.

Do we in passing all the beautiful and wondrous things produced today ever stop to think that every possible thing no matter what comes from mother nature: the softest fabrics, the finest laces, all the divine colours are products made by the brain of man, from the things around him and the earth beneath his feet. We all know this but we don't think about it! Everywhere we look are miracles! Miracles, which are no longer miracles except to the savage from the farthest ends of the earth, and we look upon him with pity for his ignorance and misunderstanding forgetting that we are all ignorant, that we all have misunderstanding!! The difference is we know his ignorance and we are none of us aware of our own.

OVERLEAF: *Koloman Moser. Loïe Fuller in the dance* The Archangel, *1902*

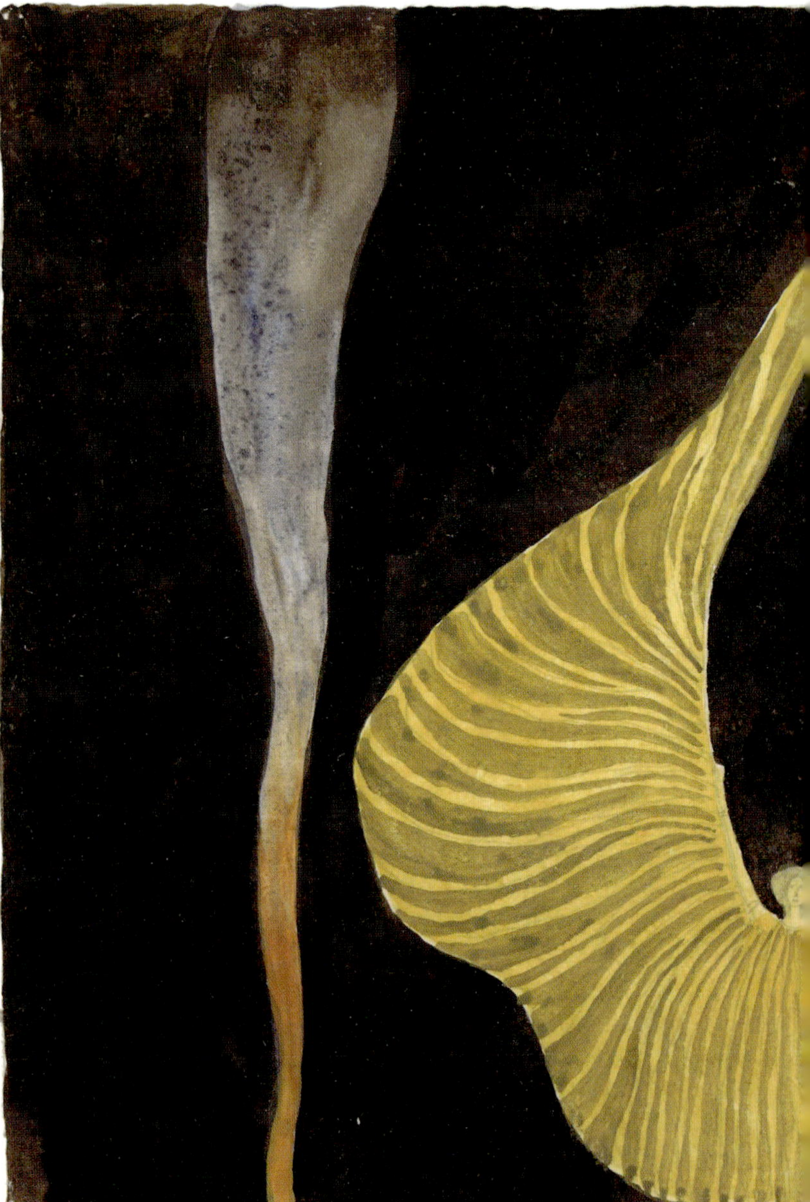

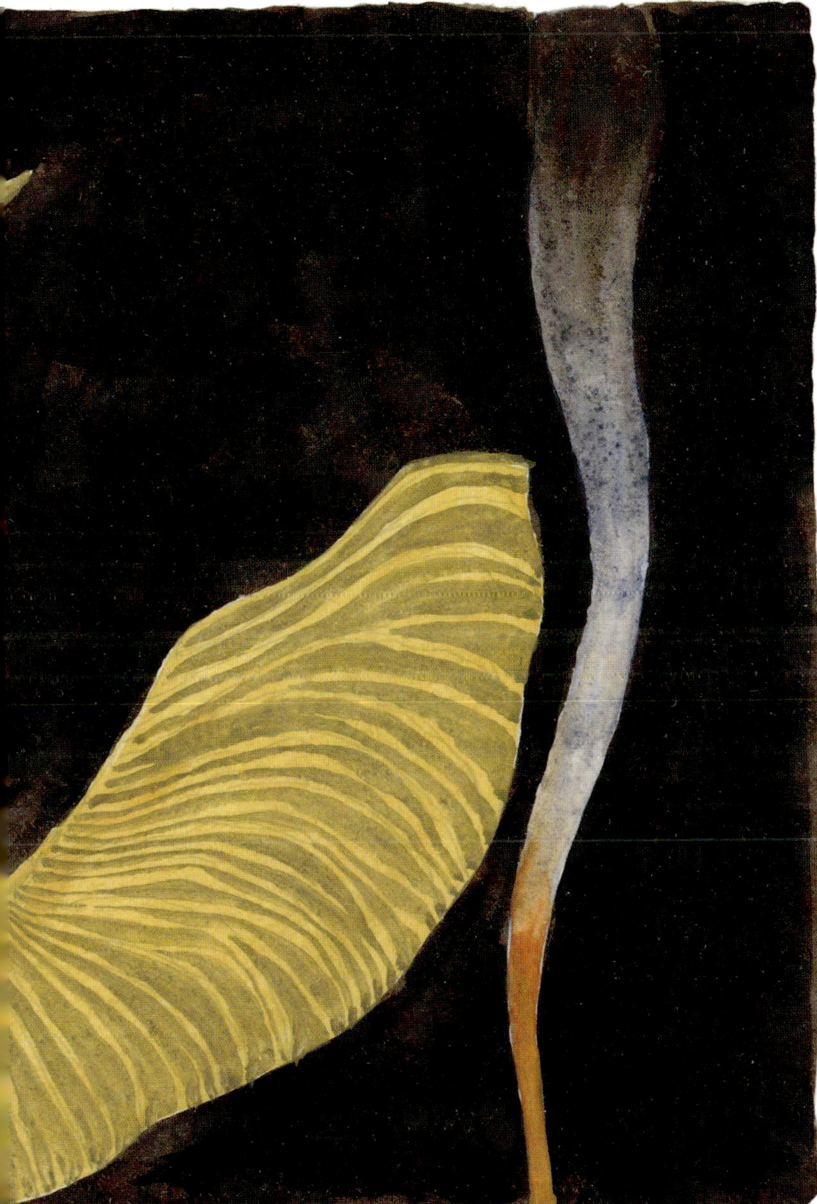

Note

If we could place far apart fifty search light or marine projectors thus & *fixed* over Paris They would all converge a half a mile over head & mixed with the city-lights – they would be impenetrable! and destruction would follow any ships audacious enough to try to enter it.

These rays also reflect from each other, thus lighting up all the atmosphere

& all the little city-lights – to fill up!!

LETTER TO CAPTAIN JOSEPH REINACH
November 1914

To conceal our city from the Zeppelins.
This can be done with enthusiastic effort.

During World War I, Loïe Fuller devoted herself to relief efforts, including frequent travels to America to raise money for war-torn France, Belgium, and Romania. Fuller's partner, Gab Bloch, worked on the front line driving ambulances that were funded by Alma Spreckels, one of the wealthiest women in America, and Fuller's most important patron. Aware of the danger of Zeppelin attacks, Fuller used her deep knowledge of stage illusions and lighting systems to devise an unprecedented military maneuver. This unsolicited letter, sent to Captain Joseph Reinach, outlines her proposal for a light defense system that will safeguard Paris. The note reproduced on page 58, included in Reinach's papers, summarizes her plan.

> *If we could place far apart fifty search lights or marine projectors thus and fixed over Paris they would all converge a half mile over head and mixed with the city lights they would be impenetrable! And destruction would follow any ship audacious enough to try to enter it.*

Thursday, November 1914
Captain Reinach

Dear Sir

With your permission I will come at 5 tomorrow to fetch you to see a light near you (in the street) which is very specially adapted to illumination. It will only take us five minutes to pass and look at it in an instant. I will bring some zinc along to show you something practical. If you are not free do not trouble because your man at the door will tell me.

Most truly yours,
Loïe Fuller

41 quai des Grand Augustin
November 1914
Captain Reinach
Dear Sir
You know I am only here with my friends who are at the front so it was not at all possible for me to find large sized paper and it was too late to find a shop open for it. So I just took what I could find at the house. It is not very nice but I know you will understand and not mind. Still worse, and much worse than that, it is written so badly—I was sleepy and could hardly keep my eyes open, and I forgot my eyeglasses in the auto. From six o'clock I was up, and so I was a little tired. If there are words

2

the City and to throw the light up in the air in order to augment the already great over head halo of a well lighted up City. — This =

instead of This →

standing above this one you can see nothing below it (Even with glasses) & from below you see all above it.

Standing here you can see nothing above, & standing above you can see all below it. Especially with big glasses which you cannot read, will you put an X so near them and I will write them over. Indeed I did many over last night, but I fear they are still difficult to read because you are not accustomed to my writing. Very sincerely yours, Loïe Fuller

Loïe Fuller
7 Nov 1914

1. I believe it would be to our advantage to be able to hear the approach of the Zeppelin.
These vibrations would be easier to catch than the vibrations of télégraphie san fil—I should think! With sensitive needles (Sounding boards—why not?) and perhaps not even the needles—stationed around Paris and equipped with the microphone (ear appliances for magnifying sound) might enable us to know, beforehand, the approach of the enemy.
2. It seems it would be an advantage to us if once over the city they could not disappear from view, these Zeppelins.
For this we must use to their fullest capacity all the city lights with reflectors directly under them (each one) to mask the city and to throw the light up in the air in order to augment the already great overhead halo of a well lighted up city.

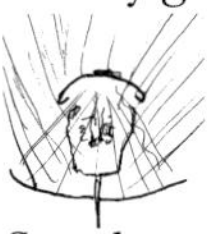

This instead of this.

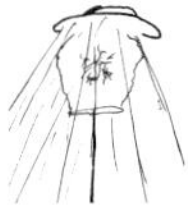

Standing above this one [upward light] you can see nothing below it (even with glasses) and from below you see all above it. Standing here [downward light] you can see nothing above, and standing above you can see all below it, especially with big glasses.

Auto mobile search lights; with storage batteries or acetylyn could be placed here & there on the river to conceal it. – simple arc lights – of the streets should also be masked! If concave, this [sketch] should be difficult to provide except with time, then [sketch] this could be easier, or even a flat thing so [sketch] – The form is not so vital as the reflection; only the concave reflectors are better. On top of monuments & houses should open arcs be established with metal underneath (also to act as reflectors.) – In order to determine the value of this, it

Automobile search lights, with storage batteries or acetylene could be placed here and there on the river to conceal it. Simple arc lights of the streets should also be masked! If concave thus would be difficult to provide except with time, then 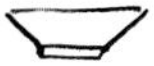 this could be easier, or even a flat thing so. The form is not so vital as the reflection, only the concave reflectors are better. On top of monuments and houses should open arcs be established with metal underneath (also to act as reflectors). In order to determine the value of this, it might be well to take (for instance) the aviating ground which is on the Seine going out towards Meudon or Bellevue, and there install over a temporarily made wooden roof which could be stuck up on poles, one lamp like this 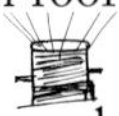 (inside this) 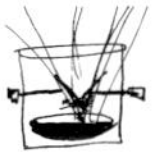 and place another one on the ground nearby, then send an aeroplane up to pass through the ray to see how high it goes up from a lamp of 40 amperes or 110 volts. I can lend these lamps for trial.

Then we could see how big an area one lamp of say 40 amperes would cover and then see how high up! and with these measurements we could figure what size lamp and what current would be required to go any distance "required." And also, the higher up, the more is the light dispersed into "atmosphere" or "light-fog" or "fog light," making it not necessary for lamps to be close together. At this same trial we should try open arc lights as well, also these same kind of lamps without the boxes. More simple yet, we can put a handle to each carbon and one man can stand on the roof and one on the ground and they make lightning by touching the carbons together and thus show a steady light or a flashing one. My electricians could do all this. It is too simple to believe. The Aviator should take strong glasses to see if he can see us below, and he should take dark spectacles to see if they protect him from the light (do they not also diminish his power to see us then—Yes). A simple, quick, and practical experiment could thus be made.—But to try the city lamps—there are 4 just at the end of the Musée du Louvre which could be masked and tried, the light is very powerful, and we could see how much 4 such lights could illuminate an aeroplane overhead. The 3rd and next vital thing of course would be to conceal the details of the city and its radius. The city (and beyond it) should be turned into a vast reflector by sending up the lights—instead of down, by having lights over the houses—open arcs with mirrors laid everywhere possible, ripolin surfaces copper zinc tin even white-oil cloth, placed irregularly—everything to shine to thus send up more and more lights into the air. Then if the enemy shoots a projector at us he only produces more and more light, his light would mingle with the lights like two lights of autos in front of each other and he could see nothing—but our aeroplanes above could see the enemy as well as the enemy could see him, and remember the enemy expects to save himself with his own projector which now is a useless member in his hands. At all times search lights are to be at hand all the same, for we must use everything. The lights of the city should be in districts, controlled so that at a signal they could snap the entire city lights off. or on. Besides—snapping lights on and off are a great thing to put the eyes out of commission. If the Germans are as cute as we have reason to believe them to be, they will paint dark their air ships and aeroplanes and they will mask themselves with light. You see this man has only to move on and snap on and off the arc. And they will snap on light only seldom and for an instant, just enough to photograph on the eyes above (which are looking down from all sides) spots for their ranges and their places for bombs.

An aeroplane which can carry 3 men can carry one man and a storage battery—remember his reflector or objective multiplies by many times his force of light. And as he can let his lamp very low down his current need not be so very great. With however a roof of "light" and the "down" underneath in shadow it is not possible to penetrate except with a greater concentrated light and no fleet of ships could provide that—and I doubt if it could be done in any case. If their search lights are so powerful that they can see something, the flashing on and off, irregularly, quickly, instead of occasionally, would render their eyes incapable of receiving an impression of any objects even though they could see them should our lights remain still and quietly burning. In case of an invasion, of course the shooting of lights as irregularly and as rapidly as possible would be our guns, incapacitating our enemy. If he used glasses darkened it is true he would see less light but he would also be all the more incapacitated for seeing us, or our city below the lights and our realm of light above would make it possible for our ships to keep them always in view. If darkness would be good for our aeroplanes, it would certainly be better for the Zeppelins. So light is preferable, for then we could also shoot from below.

With the "light" defense system the signals of spies through glass roofs would become valueless. Also any information of localities, also information as to where our marine projectors would be placed. At present our marine projectors (if we let our eye follow down to the end of the ray) mark very plainly where they are, don't they? Our marine projectors illuminate too

As most of our appliances for projection came from Germany they are well up in the use of all rays of light. — with a little hand mirror it is possible to send the ray where you like, this

mirror

a lamp can be laid on a table or hung on a wire.

As I said — three things are vital

To hear the Zeppelins —
To see the Zeppelins —
To conceal our city — from the Zeppelin
This can be done with enthusiastic effort

much a small spot. This 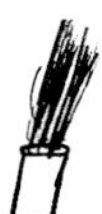 instead of this. 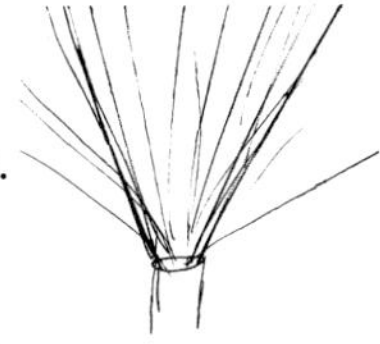

It is not necessary to have such a thick light because one wants to see more much more.

To illuminate an object, and to see it, are two different things, it seems to me we want to see it! Therefore; more illuminated atmosphere over us, and with us well masked underneath and less illuminated objects is the thing.

Spies will of course find all information, but even knowing it the enemy can do nothing, but throw bombs anywhere which I hope with our sounding board signals and our lighted sky to show their whereabouts and our own Airmen, there would not be many bombs thrown down on us.

As most of our appliances for projection come from Germany they are well up in the use of all rays of light. With a little hand mirror it is possible to send the ray where you like, thus

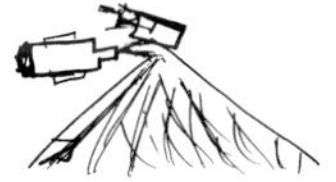

A lamp can be laid on a table or hung on a wire.

As I said—three things are vital
To hear the Zeppelins—
To see the Zeppelins—
&
To conceal our city from the Zeppelins.
This can be done with enthusiastic effort.

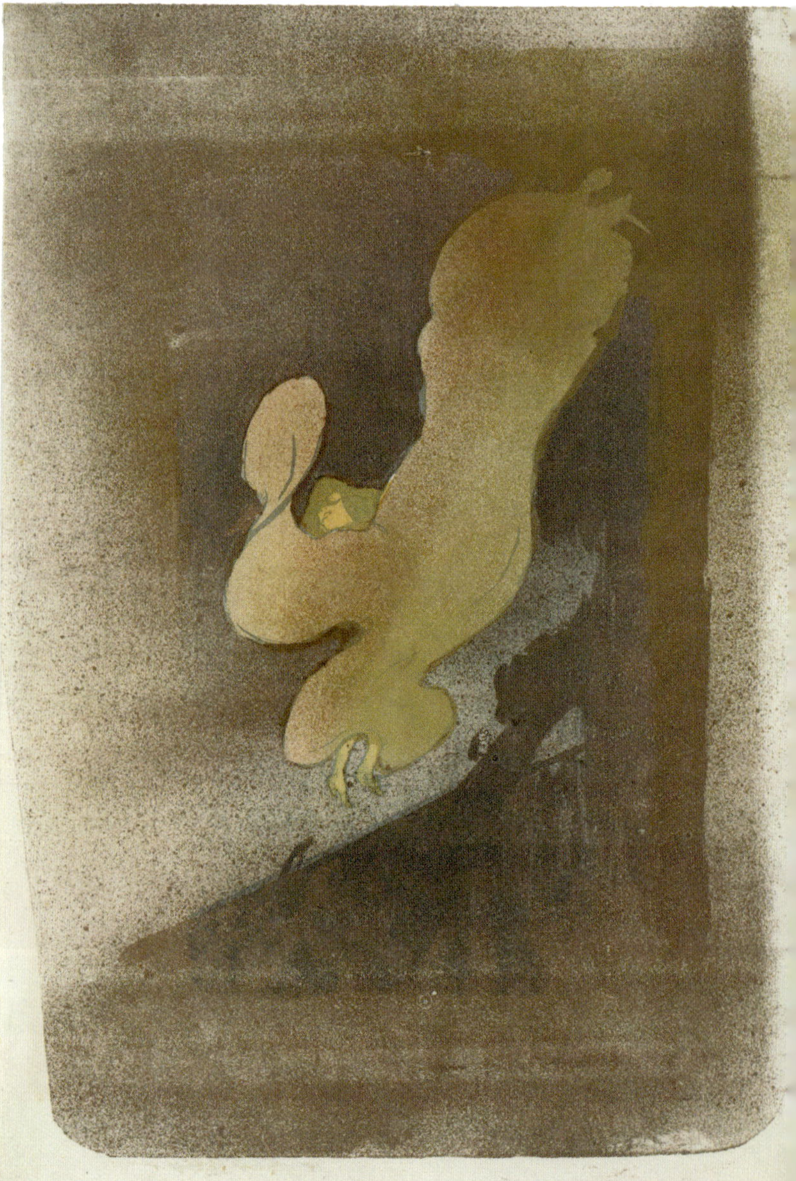

PRELUDE TO LIGHT
ca. 1911

We will begin with colour.

This fragmentary text appears in a notebook that also contains drafts of the "Lecture on Radium" and "How I Became Interested in Radio-Active Matter." The text ends with an incomplete sentence on the final page, where Fuller adds one more note: "Light—Movement and the power of colour in movement. We will begin with colour." The rest of the notebook is blank.

PAGE 70: *Henri de Toulouse-Lautrec,* Miss Loïe Fuller, *1893*

There is nothing more difficult to describe in words than the sight of a passing, ever moving, shifting, vaporous "something" which takes indefinite form and colour without tangible shape.

The elements may be likened to this.

The sensation which they convey cannot be described in any language, they must be seen and felt to be understood.

Fire, storm, wind, waves, smoke, sky—what are they? Only our senses can answer that.

Instinct and sensation, those elements within us which enable us to understand and know that there exists things not made of so-called material—and yet they are the greatest real material in so much as they direct our knowledge on or to the highest point of man's education—"Understanding."

I asked an old very old gentleman who was a noted geologist if he had written all his knowledge down. He replied, I can't. How can I tell you about the things I cannot explain. Can you describe how you know the walk, even normal walk, of a person, their voice or handwriting? I can write down everything but the most important part—Judgement—I cannot tell that.

Yes instinct and understanding are nature's real directors, using the brain and intelligence to indicate to us their real value.

Thus, a living moving mass of colour or colours must be seen and felt to be known and understood.

So unliving a thing is it, that it exists only in itself and can no more be copied than wind or storm are copied on the canvas. Things in a storm, yes, but that gives us no idea of what the storm or wind would make us feel.

Could any canvas make us feel heat or cold, or suggest it if we had never felt them? No, one must come in contact with the elements of nature whether they are interior or exterior, in order to realize them. We must sense things to know them and real education is to learn what the "sense" means! The brain, the business of the brain should be to learn it, not to direct it, as is the usual custom, a fact which might provide our schools with a new line of study—the use of instinct.

In using, producing and arranging rays of light and colour, our knowledge is about where it was with regard to music before there was any. For on reflection we must know that nature's sounds and noises were the first and only music before the brain of man discovered that he could produce and govern harmony out of sound.

So are we beginning to see that we can direct to great use and beauty the decomposition of rays of light. We know that colour is in fact a condition of light rays and waves, but to play upon it for orchestrations of harmony to the eye—as music is to the ear—seems far, far away into the unknown world, and so it is but since Science has produced powerful lenses that can magnify and bring to our imperfect sight a perfect sight, we are enabled to see and learn the wonders of decaying light, in the microscopic world, a world not yet made use of for beauty's sake but rather for different materials of which . . . [end of text]

Light—Movement and the power of colour in movement. We will begin with colour.

(No Model.)

M. L. FULLER.

GARMENT FOR DANCERS.

No. 518,347. Patented Apr. 17, 1894.

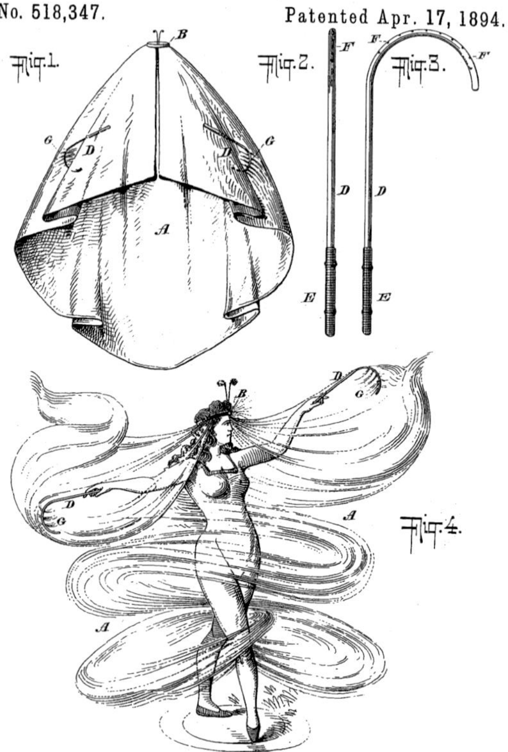

WITNESSES:

Gustav Dieterich

H. B. Brownell

INVENTOR

Marie Louise Fuller.

BY

R. C. Mitchell

ATTORNEY.

THE NATIONAL LITHOGRAPHING COMPANY, WASHINGTON, D. C.

PATENTS BY LOÏE FULLER

GARMENT FOR DANCERS
APRIL 17, 1894. PATENT NO. 518,347

MECHANISM FOR THE PRODUCTION OF STAGE EFFECTS
JANUARY 23, 1894. PATENT NO. 513,102

THEATRICAL STAGE MECHANISM
JANUARY 29, 1895. PATENT NO. 533,167

UNITED STATES PATENT OFFICE.

MARIE LOUISE FULLER, OF NEW YORK, N. Y.

MECHANISM FOR THE PRODUCTION OF STAGE EFFECTS.

SPECIFICATION forming part of Letters Patent No. 513,102, dated January 23, 1894.

Application filed October 12, 1893. Serial No. 488,005. (No model.) Patented in France April 8, 1893, No. 227,105, and in England May 24, 1893, No. 10,301.

To all whom it may concern:

Be it known that I, MARIE LOUISE FULLER, a resident of the city, county, and State of New York, have invented certain new and useful Improvements in Mechanism for the Production of Stage Effects, particularly of an illusionary character, (for which I have obtained patents in Great Britain, No. 10,301, dated May 24, 1893, and in France, No. 227,105, dated April 8, 1893,) of which the following is a specification.

Figure 1 is a perspective view of a theatrical stage setting, partially in section, to show the illuminating foyer below the stage. Fig. 2 is a vertical section of a portion of my invention by which an illusionary effect is produced. Fig. 3 shows a lighting device to be used in connection with my invention, and Fig. 4 shows a modification of one of the details of my invention.

A is an ordinary stage flooring.

B is a false flooring of glass, or wood and glass, in which latter case the wooden flooring may be pierced with a number of holes of any suitable size or shape in which may be embedded thick and transparent plaques of glass or lenses D D. Beneath each of these plaques, I place electric or calcium lights E so disposed that the rays thereof will be projected upward by any suitable reflector through the plaques or lenses D D. At one or more points in the false staging I can, if desirable, place one or more pedestals F, each of which supports at its upper end a suitable plaque G. This pedestal may be of plain or fanciful form and by preference it is hollow and made of glass for the purpose hereinafter described. The plaque G, by preference, is made of glass and is preferably silvered on its lower side, outside of the pedestal, the back of the silvering being blackened for the purpose hereinafter described. The center of the plaque, or that portion directly above the pedestal, is preferably transparent.

By preference the inner side of the pedestal F, when made from glass, is silvered; the sides of the pedestal when it is silvered on the inner side diverge from their lower end toward the top, as shown in Fig. 4 so as to reflect toward the audience the stage flooring. Underneath each pedestal, in the illuminating foyer, between the floor A and false flooring B, is placed an electric or calcium light which reflects the rays of light directly upward through the pedestal, and through the central transparent portion of the glass plaque.

In producing the illusionary effect, it is desirable that the stage B upon which the pedestal F rests, or the lower floor A, the background H, wings I and flies J may be of a solid color, preferably black, in which case it will be readily seen that the pedestal F will reflect the black flooring inasmuch as the pedestal, when made of looking-glass, diverges from the bottom toward the top, thereby affording to the dancer, a support that is invisible to the audience. It will now be seen that the figure of the dancer clothed, preferably, in white or in some color sufficiently contrasting with the color of the scene, standing on a plaque in the floor, or on a plaque on the top of the pedestal, will appear to be mysteriously suspended in the air. By turning on the light from below and by illuminating, if desirable, from the wings of the stage by projected light, the figure will, as before stated, be apparently suspended in the air. Graceful evolutions may be performed without marring the illusion. The rays of light coming up through the center of the plaque upon which the figure stands, will illuminate the garment of the dancer which garment will be reflected by means of the looking-glass plaque G toward the observer, exactly as the figure of the person and the garment moves upon the plaque, the remainder of the plaque, (providing the figure does not entirely cover it) reflecting merely the darker back-ground or flies to the rear and above.

If it is desirable that dancers perform on the floor around the pedestal, the said dancers may step directly over the various plaques inserted in the floor of the stage through which lights are projected and the said dancers will seem to dance in a luminous aureola. If desirable, the whole stage floor may be made of glass, in which case the reflectors of the lights are pointed so as to reflect none of the rays toward the audience, in which case all the dancers will seem to be suspended in space by reason of the transparent flooring. It is obvious that when the entire

(No Model.)

M. L. FULLER.

MECHANISM FOR THE PRODUCTION OF STAGE EFFECTS.

No. 513,102. Patented Jan. 23, 1894.

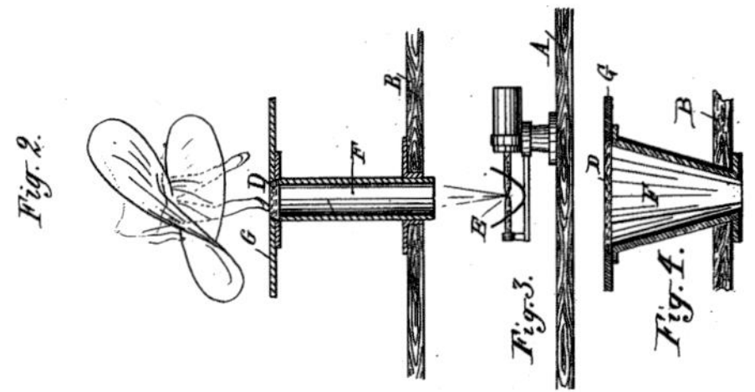

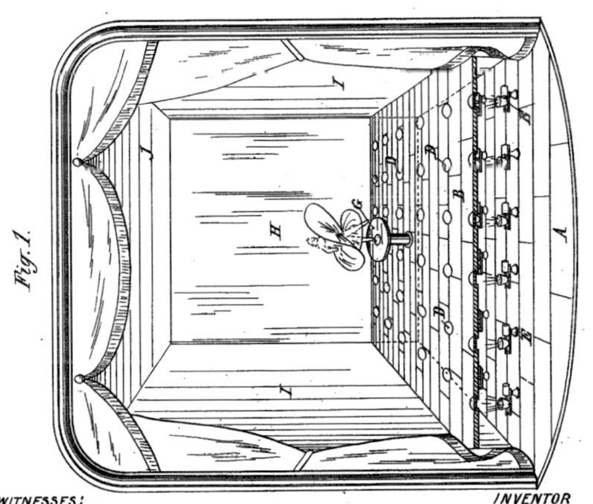

WITNESSES:

H. B. Brownell

H. W. Skinner

INVENTOR

Marie Louise Fuller.

BY

R. C. Mitchell.

ATTORNEY.

THE NATIONAL LITHOGRAPHING COMPANY, WASHINGTON, D. C.

flooring is of glass, the illuminating foyer must be of the same color as the drapery of the stage setting.

The object of coloring the lower side of the plaque supported on the pedestal, as described, is to prevent the audience from detecting the presence of the plaque from below. If desirable the false floor B may be discarded and the plaques placed in the ordinary stage floor.

Having thus described my invention, what I claim, and desire to secure by Letters Patent, is—

1. In a stage mechanism, a looking-glass pedestal F the sides of which pedestal diverge from the bottom toward the top, supported on the stage, said pedestal supporting a plaque G all adapted to produce the illusion, substantially as and for the purpose specified.

2. In a stage mechanism a floor upon which rests a hollow pedestal of looking-glass supporting a glass platform or plaque G which is transparent above the pedestal and silvered on all the lower side except that portion covered by the pedestal, substantially as and for the purpose specified.

3. In a stage mechanism, a hollow looking-glass pedestal F supported on the floor around a transparent plaque in the floor, the sides of the pedestal diverging from the bottom toward the top, upon the upper end of which pedestal is supported a looking-glass plaque transparent directly above the pedestal, the stage decoration being of a plain color in combination with the illuminating device E, substantially as and for the purpose specified.

MARIE LOUISE FULLER.

Witnesses:
R. C. MITCHELL,
H. B. BROWNELL.

(No Model.)

M. L. FULLER.

THEATRICAL STAGE MECHANISM.

No. 533,167. Patented Jan. 29, 1895.

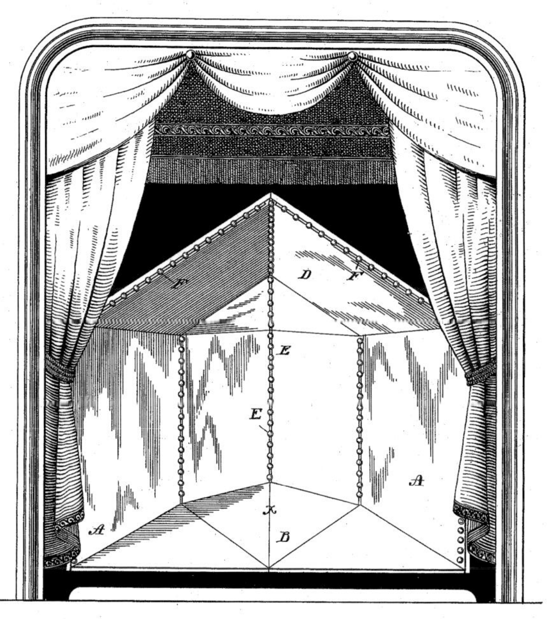

WITNESSES:

Frank S. Ober.

Hettie W. Skinner

INVENTOR

Marie Louise Fuller.

BY

R. C. Mitchell.

ATTORNEY

UNITED STATES PATENT OFFICE.

MARIE LOUISE FULLER, OF NEW YORK, N. Y.

THEATRICAL STAGE MECHANISM.

SPECIFICATION forming part of Letters Patent No. 533,167, dated January 29, 1895.

Application filed March 27, 1894. Serial No. 505,258. (No model.) Patented in France January 13, 1893, No. 227,106, and in England May 23, 1893, No. 10,221.

To all whom it may concern:

Be it known that I, MARIE LOUISE FULLER, a citizen of the United States, residing at New York, county of New York, and State of New York, have invented a new and useful Theatrical Stage Mechanism for Producing Illusionary Effects, (for which I have obtained a French patent, No. 227,106, dated January 13, 1893, and delivered April 8, 1893, and a patent in Great Britain, No. 10,221, dated May 23, 1893,) of which the following is a specification.

My invention consists in constructing a new and useful theatrical stage mechanism to be used in the production of a dance, said mechanism having the effect of multiplying the reflections of one or more dancers performing in front of the said mechanism, thereby producing to the eye of the spectator an illusionary effect.

In a modification of my invention, I provide a series of vertical lights which give the effect of many pillars of fire around, and among which the figures are represented as dancing.

The object of my invention is to produce an illusionary effect to the eye of the spectator.

My invention is illustrated by the accompanying drawing, in which the single figure therein illustrates a front elevation of my invention.

A A are plates of glass, silvered on the back so as to produce reflecting surfaces. These reflecting surfaces are set adjacent to one another and at an angle to each other, the vertical edge or edges of each of said planes being placed adjacent to each other, as shown. Two or more reflecting planes may be placed adjacent to one another and several angles may be formed thereby for the purpose of forming a background of semi-polygonal shape. A mirror flooring B may be added within the space partially surrounded by the background. A mirror ceiling D may also be added. When it is desirable to use a mirror ceiling, by preference, I place the plates which form the same at an incline, so that spectators, who are witnessing the illusion from a plane higher than the plane of the flooring B, will not be prevented from seeing the entire background. At each vertical juncture of the reflecting surfaces that form an angle, by preference, I provide a vertical series of lights E E. Back of and around the said mechanism by preference I provide a black setting. Around the front edge of the mechanism, by preference, I provide a series of lights F F, which lights may be shielded from the eyes of the spectator, if desired.

In operation, it will be clearly seen that when the dancer comes in front of the mechanism and stands upon the dancing flooring B, which forms a part of this invention, at a point, for illustration, lettered X, the reflections caused by the mirrors A A will make it appear to the spectator that there are many dancers performing upon a large stage. The vertical columns of light E will likewise be multiplied, and the dancers will be seen going through the various evolutions of the dance among these various pillars of fire, thereby producing an extraordinarily beautiful effect. By preference, I use incandescent electric lights, and, by providing various colored bulbs for the same, it is possible to instantly change the tone of the scene from one color to another, and by combining the various primaries other colors are produced. When the mirror flooring B and the inclined mirror ceiling D are utilized in connection with the vertical reflecting surfaces A A, and all in combination with the rows of lights, the reflections of the dancer will be increased many fold and the illusion made more mystifying and complete. To heighten the effect, all the lights in the auditorium may be turned off and the only lights used should be those described in the specification, with possibly the addition of a "lime" or calcium light applied from behind the wings of the stage.

Several vertical rows of light E E are shown and these may be used independently or jointly during the progress of the dance, as desired. It is furthermore desirable, in order that the number of figures may be compounded, that the mirrors be so set as to reflect into one another. It would be desirable, therefore, when only two plates are used to set the said plates at an acute angle to each other, thereby making it appear that there are five persons dancing while, in reality, there is only one dancer performing.

Having thus described my invention, what I claim, and desire to secure by Letters Patent, is—

The combination in a theatrical stage mechanism of a background of reflecting surfaces placed adjacent to and at angles to each other, with a mirror flooring and mirror ceiling, said mirror ceiling being inclined upward toward the front of the stage, and with a vertical column of lights arranged at the angles formed by the junction of the background mirrors, substantially as described.

MARIE LOUISE FULLER.

Witnesses:
CLYDE SHROPSHIRE,
HENRY PEARTREE.

Date June 11. 1916

Name [illegible]

Current, Richard Nelson, and Marcia Ewing Current. *Loïe Fuller: Goddess of Light*. Northeastern University Press, 1997.

Dillon, Brian. *Affinities: On Art and Fascination*. New York Review of Books, 2023.

Fuller, Loïe. *Fifteen Years of a Dancer's Life: With Some Account of Her Distinguished Friends*. Introduction by Anatole France. Herbert Jenkins Limited, 1913 (originally published in French as *Quinze ans de ma vie*. Paris: Félix-Juven, 1908).

Garelick, Rhonda K. *Electric Salome: Loïe Fuller's Performance of Modernism*. Princeton University Press, 2007.

Goldsmith, Barbara. *Obsessive Genius: The Inner World of Marie Curie*. Atlas Books/W. W. Norton, 2005.

Gunning, Tom. "Loïe Fuller and the Art of Motion: Body, Light, Electricity and the Origins of Cinema." In *Camera Obscura, Camera Lucida: Essays in Honor of Annette Michelson*, edited by Richard Allen and Malcolm Turvey. Amsterdam University Press, 2003.

Herrera Gómez, Aurora, ed. *Body Stages: The Metamorphosis of Loïe Fuller*. Skira, 2014. Published in conjunction with an exhibition of the same title, presented at La Casa Encendida, Madrid, February 6–May 4, 2014.

Kermode, Frank. "Loïe Fuller and the Dance Before Diaghilev." *Theatre Arts*, September 1962.

PAGE 86: *Loïe Fuller's signature from her autograph album "The Ghosts of Friends." This store-bought album was designed for the collection of signatures, which, when folded, create suggestive images. Fuller used the album between 1900 and 1928, and it contains signatures from many of her illustrious friends and acquaintances, including Camille Flammarion, Auguste Rodin, and Rudolph Valentino.*

THE EVENING SUN, BALTIMORE, SATURDAY, MAY 5, 1928

No Americans Among 10 Foremost Women: Only One From U. S. Mentioned in Voting— Loie Fuller, Dancer, Who Died in Paris.

Paris, May 5 (AP).—The ten most important women the world has produced are all Europeans in the majority judgment of more than a million French newspaper readers who voted on the subject.

Only one American was mentioned—Loïe Fuller, the dancer, who died in Paris a few months ago.

Mme. Curie, famed for research with radium, led the list, which was compiled by the newspaper *Quotidien*.

Second in order of popularity was Sarah Bernhardt, followed by Nurse Edith Cavell, George Sand, Evangeline Booth, Louise Michel, the "red virgin" of the Commune; the Countess of Noailles, French poetess; Mme. Severine, a radical journalist; Suzanne Lenglen and Mme. de Stael, who mixed politics with tea parties in Napoleonic days.

Besides Loïe Fuller, mention was made of Emmeline Pankhurst, the suffragist, and George Eliot, the novelist.

The Evening Sun, *Baltimore, Saturday, May 5, 1928*

NO AMERICANS AMONG 10 FOREMOST WOMEN

Only One From U. S. Mentioned In Voting—Loie Fuller, Dancer, Who Died In Paris.

Paris, May 5 (AP).—The ten most important women the world has produced are all Europeans in the majority judgment of more than a million French newspaper readers who voted on the subject.

Only one American was mentioned—Loie Fuller, the dancer, who died in Paris a few months ago.

Mme. Curie, famed for research with radium, led the list, which was compiled by the newspaper *Quotidien.*

Second in order of popularity was Sarah Bernhardt, followed by Nurse Edith Cavell, George Sand, Evangeline Booth, Louise Michel, the "red virgin" of the Commune; the Countess of Noailles, French poetess; Mme. Severine, a radical journalist; Suzanne Lenglen and Mme de Stael, who mixed politics with tea parties in Napoleonic days.

Besides Loie Fuller, mention was made of Emmeline Pankhurst, the suffragist, and George Eliot, the novelist.

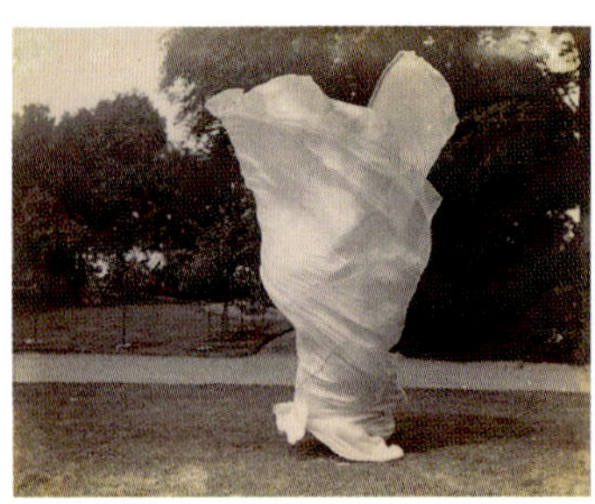

FRONT COVER
Samuel Joshua Beckett. [Loïe Fuller Dancing], ca. 1900

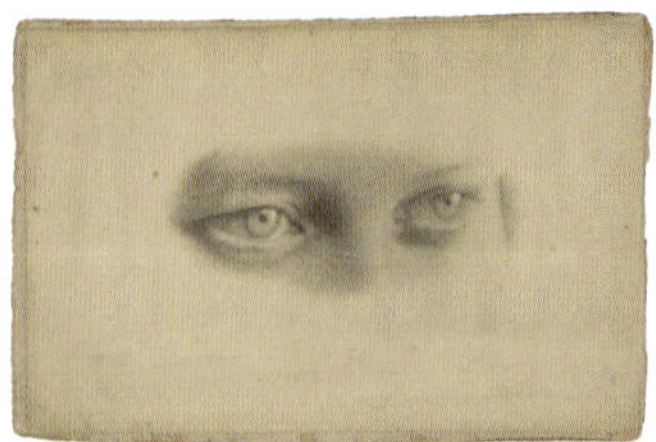

BACK COVER
Loïe Fuller, Dance of the Eyes, *1907*

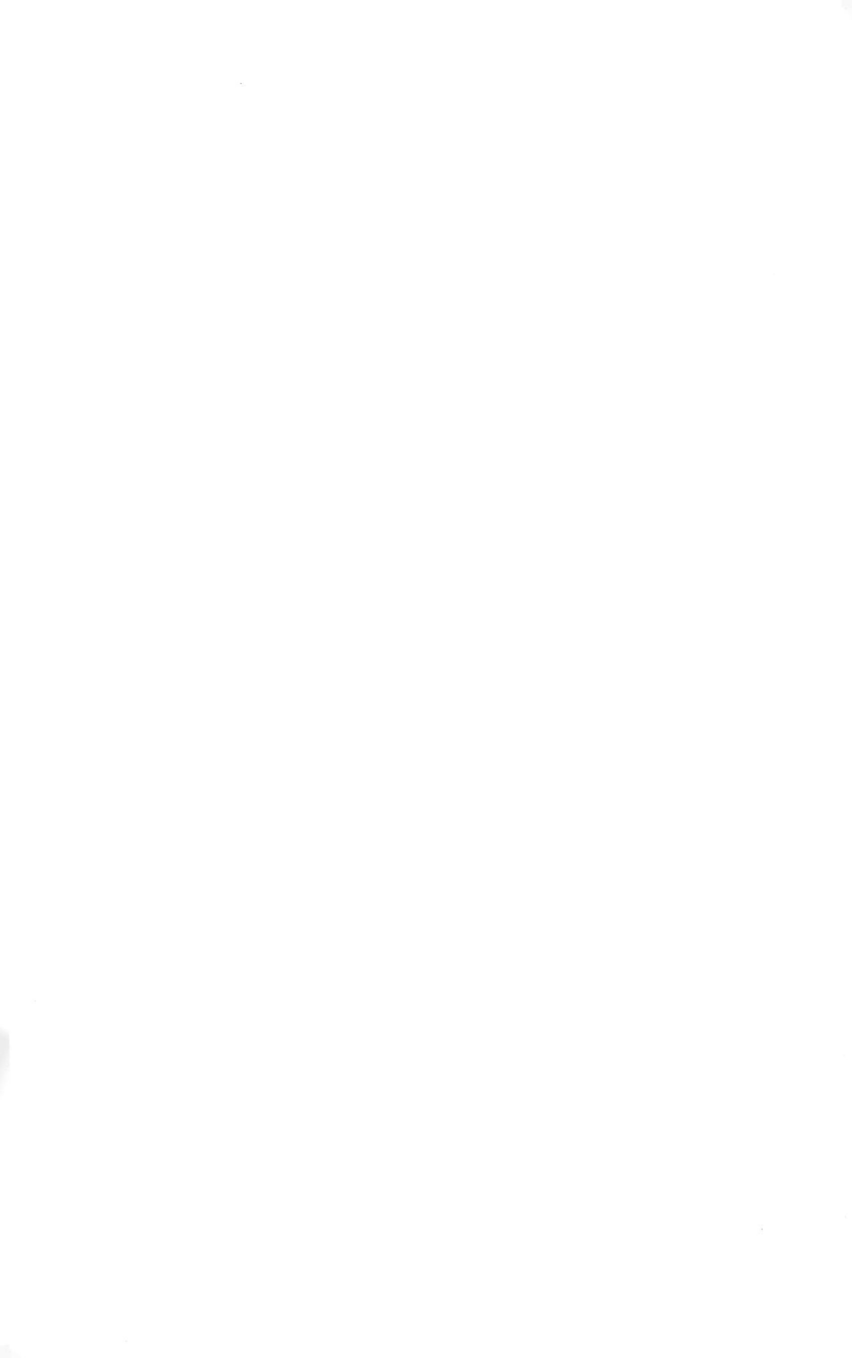